Graphic
Trisection
of an Arbitrary Angle

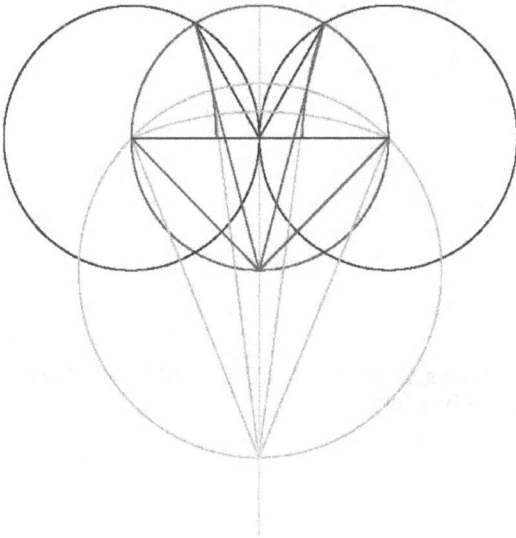

The *Solution* to the *Impossible* Problem

by
Harold Florentino LATORTUE, PhD

Library of Congress Control Number: 2017905297
FLatortue, Brooklyn, NY

FLatortue LLC
Brooklyn, NY 11225
FLatortueMethod@gmail.com

Printed in the United States of America

IN MEMORIAM

Of My Mother,

The 'Lady' Olga Latortue

and

Of My Father,

The 'Leopard' Fresnel Casseus

Both of you refused to teach me how to accept Limits.

You really made my life

LIMITLESS

SPECIAL THOUGHTS

Of late

Iseline Lamarre Calixte

You impacted my world.

DEDICATION

To my son

Didier Lawrence Latortue

and

to my daughter

Cryst-Ena Latortue

I am blessed with the best kids ever.

I am proud of you.

SPECIAL THANKS

To

Caroline

You are really the 'Pest'.

'Ji'

Hane, you are **Vital.**

Nobody can prove that a problem is impossible to resolve.

One can only demonstrate that they cannot provide

a solution.

Florentino Latortue

I am not discovering new mathematic theories

but

new ways to think about mathematic solutions.

Florentino Latortue

Preface

The FLatortue Method to trisect an arbitrary angle α, is addressed to everyone who has some interest in mathematics and geometry, from the general public to the midlevel secondary students attending a geometry class, to the high school teachers, to the college students and math professors at graduate levels in a university. The Flatortue Method of an arbitrary angle trisection provides the necessary basic knowledge (to trisect an arbitrary angle using a compass and a straight edge) that was lacking in the fields of mathematics and geometry studies for centuries. The FLatortue Method declassifies trisection from an 'impossible' problem to common knowledge. The FLatortue Method opens the doors to solving the problem of dividing an arbitrary angle α into 'n' equal angles where 'n' is a prime number (n equals to 3, 5, 7, 11, etc.).

In this book, the simple steps to perform the trisection of an arbitrary angle are presented along with the algebraic analysis that shows why the FLatortue Method is mathematically justified.

Harold Florentino LATORTUE, PhD

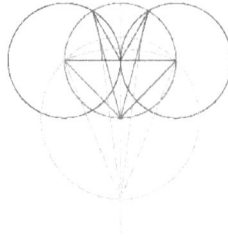

Graphic Trisection of an Arbitrary Angle α
The FLatortue Method

Introduction:

Nobody can prove that a problem is impossible to resolve. One can only demonstrate that they cannot provide a solution. However, trying to or claiming to be able to resolve a problem classified as 'impossible', by all in the field since the dawn of science, is most of the time labeled as 'very presumptuous'. In fact, this explains why such a person, is described by this quote from **https://en.wikipedia.org/wiki/Angle_trisection**:

" Because it is defined in simple terms, but complex to prove unsolvable, the problem of angle trisection is a frequent subject of pseudomathematical attempts at solution by naive enthusiasts. These "solutions" often involve mistaken interpretations of the rules, or are simply incorrect. "

Trisecting of an arbitrary angle α using just a straightedge and a caliper 'compass' is classified as one of the impossible geometric problems to solve up to today. Pierre Wantzel, in 1837, published a study where he concluded that the trisection of an arbitrary angle α is generally impossible to do using a straight edge and a compass except for a few specifics values of α such as 180^0, 90^0, etc. Today, most mathematicians agree with this statement.

However, following the same logic, a scientist from prior to 1492, would have concluded that the earth was flat and would have easily provided mathematical proof of his/her declaration. But Christopher Columbus kept his beliefs of the contrary. His 'presumptuous' character made him a very famous man whose name is still well known all over the world five hundred and twenty-five years later.

My guess is that: *'It is harder to convince fellow human beings that you can resolve their 'impossible problems' than to find the solutions.'*

Underwood Dudley wrote:

> *'A trisector is a person who has, he thinks, succeeded in dividing any angle into three equal parts using straightedge and compass alone. He comes when he sends you his trisection in the mail and asks your opinion, or (worse) calls you to discuss his work, or (worse still) shows up in person. You may think that the problem of how to deal with trisectors is not an important one; I intend to show that it is'.*

Fortunately for humanity, the King and the Queen of Spain did not agree with Underwood Dudley's approach. When Columbus showed up in person to solicit their help for his journey that led to the discovery of the new world, they listened.

I have decided to be as arrogant as Christopher Columbus to set the objective of this study as to show that the statement of Pierre Wantzel (1837) is false. I am proving in this book that there is a simple and quite easy way to resolve the 'impossible' problem of 'Trisecting an Arbitrary angle α using just a compass and a straightedge. This study provides the methodology to achieve such solution for any angle from 0° to 360° using exactly what is required in the Greek statement of the trisection problem.

Problem statement:

Among several options someone can, to define the trisection problem, quote the words of this Internet site, (https://terrytao.wordpress.com/2011/08/10/a-geometric-proof-of-the-impossibility-of-angle-trisection-by-straightedge-and-compass/)

One of the most well-known problems from ancient Greek mathematics was that of trisecting an angle by straightedge and compass, which was eventually proven impossible in 1837 by Pierre Wantzel, using methods from Galois theory.

Formally, one can set up the problem as follows. Define a configuration *to be a finite collection C of points, lines, and circles in the Euclidean plane. Define a* construction step *to be one of the following operations to enlarge the collection C:*

- *(Straightedge) Given two distinct points A and B in C, form the line AB that connects A and B, and add it to C.*
- *(Compass) Given two distinct points A and B in C, and given a third point O in C (which may or may not equal A or B), form the circle with center O and radius equal to the length |AB| of the line segment joining A and B, and add it to C.*

- *(Intersection) Given two distinct curves γ and γ' in C (thus γ is either a line or a circle in C, and similarly for γ'), select a point P that is common to both γ and γ' (there are at most two such points), and add it to C.*

We say that a point, line, or circle is constructible by straightedge and compass from a configuration C if it can be obtained from C after applying a finite number of construction steps.

The graphic solution:

1 - For a given angle $\alpha°$ of with apex A,

What am I doing?

You are starting the process to divide the given angle $\alpha°$ into three (3) equal angles (trisection) using only a straightedge and a compass. The angle $\alpha°$ is arbitrary. Its value or size is not known. For this study, we do our analysis when the angle $\alpha°$ is between $0°$ and $180°$. For angle $\alpha°$ larger than $180°$, one can (among many available options) work on the angle $\phi° = 360° - \alpha°$, apply the FLatortue method on $\phi°$. Then subtract the result ($\phi°/3$) from a $120°$ angle to get the solutions for the trisection of the angle $\alpha°$ larger than $180°$.

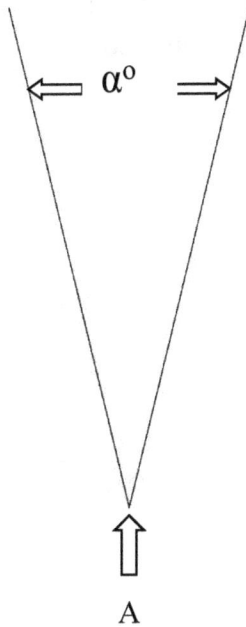

Figure 1 - Given angle α° of with apex A

Graphic Trisection of an Arbitrary Angle α
by Harold Florentino LATORTUE, PhD

What am I doing?

You are defining the points B and C, which are the key points for the trisection. The segment BC is the diameter of the trigonometric circle that you will construct in the next steps. Segment BC is the cosines axis. However what BC represents is not important for the graphic solution of trisection. You must just keep in mind the locations of both B and C.

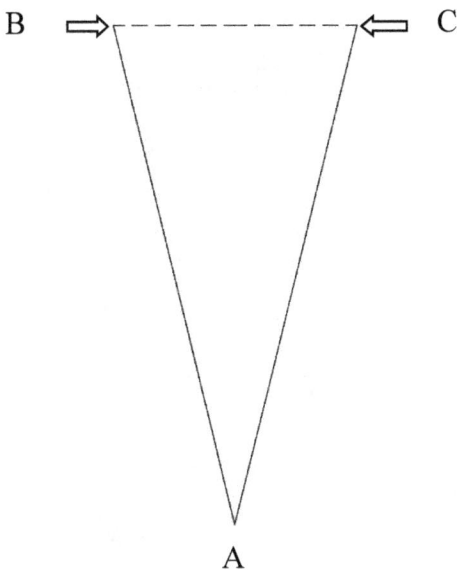

Figure 2 - Construct an isosceles triangle ABC

Graphic Trisection of an Arbitrary Angle α
by Harold Florentino LATORTUE, PhD

3 - Construct the bisection of the angle α^0 with apex A.

What am I doing?

You are dividing the angle BAC in two (2) equal angles. The bisection line defines the sinus axis of the trigonometric circle that you will construct. Knowing that the bisection line represents the sinus axis is not important. But, the bisection line is the most important line of the graphic method of the trisection.

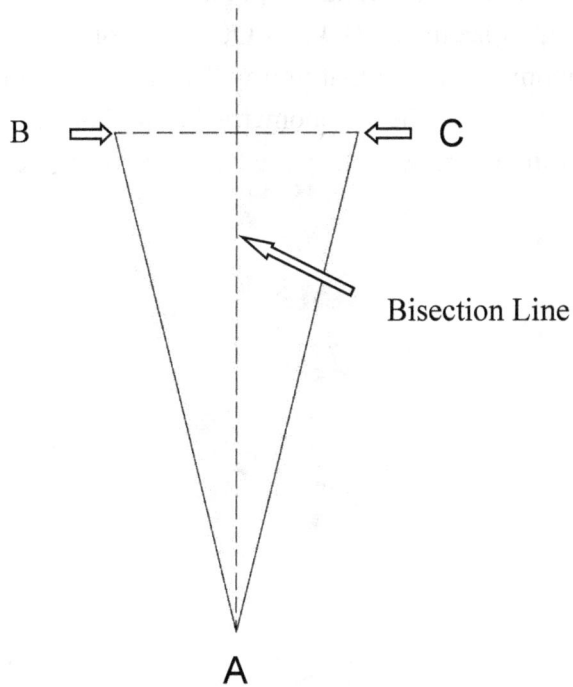

Figure 3 - Construct the bisection of the angle α°

Graphic Trisection of an Arbitrary Angle α
by Harold Florentino LATORTUE, PhD

4 - Mark the point O, intersection of the bisection line and the segment BC.

What am I doing?

You are defining the point that divides the segment BC into two equal segments BO and OC. The point O is the center of the trigonometric circle that you will construct. BO and OC are equal to the radius of the trigonometric circle. For the graphic method, it is not important what they are but where they are.

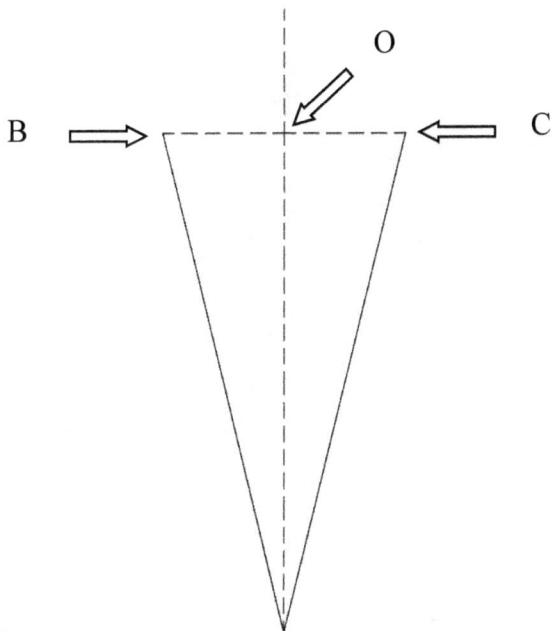

Figure 4 - Mark the point O, intersection of the bisection line and the segment BC

Graphic Trisection of an Arbitrary Angle α
by Harold Florentino LATORTUE, PhD

5 - Draw circle C_1 with center O and radius equals OB.

What am I doing?

You are drawing the trigonometric circle mentioned above.

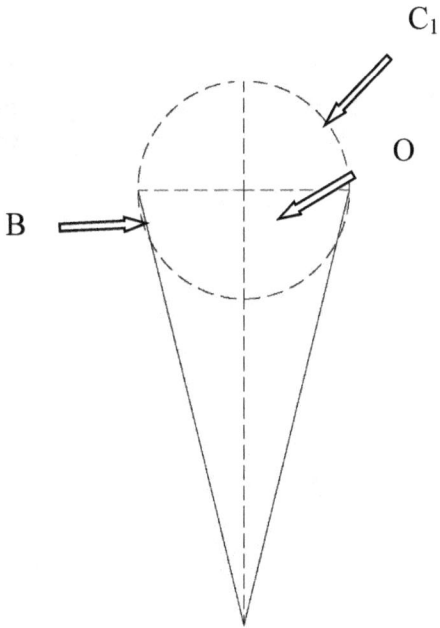

Figure 5 - Draw circle C_1 with center O and radius equals OB.

What am I doing?

You are drawing Circle C_2 which, in addition with Circle C_1 and the next following circle, gives you all you need to trisect an angle α^o equals to 180° into three equal angles of 60°.

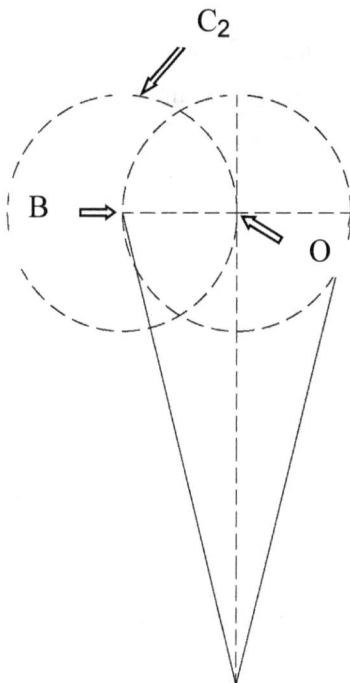

Figure 6 - Draw circle C_2 with center B and radius equals BO.

7 - Draw circle C_3 with center C and radius equals CO.

What am I doing?

You are drawing Circle C_3, which with circle C_1 and circle C_2 constitutes the basis for trisecting an 180° angle into three equal angles of 60°.

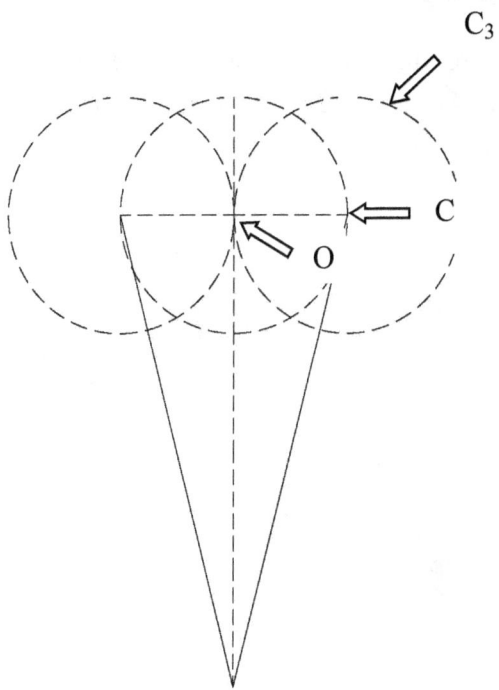

Figure 7 - Draw circle C_3 with center C and radius equals CO.

What am I doing?

You are finding the first point solution for trisecting an angle of 180°. The angle BOD is that solution and its value is 60°.

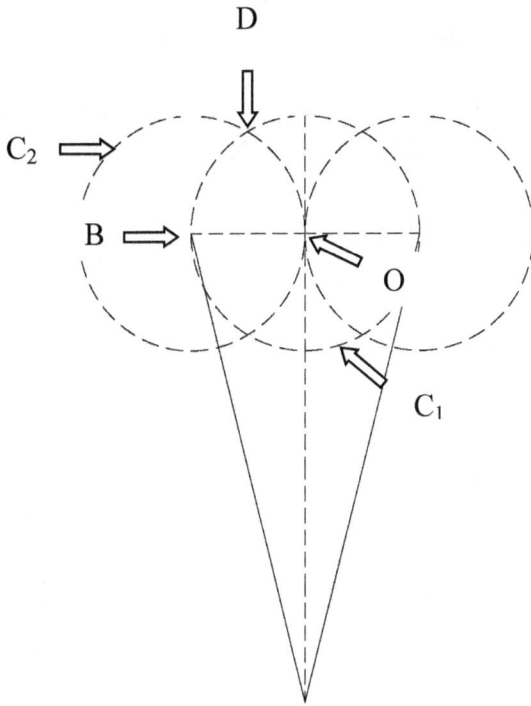

Figure 8 - Mark D, the top intersection of Circle C_1 and C_2.

What am I doing?

You are finding the second point solution for trisecting an angle of 180°. The angles DOE and EOC are the other two angles that solve the trisection of an 180° angle. Their values are 60° each.

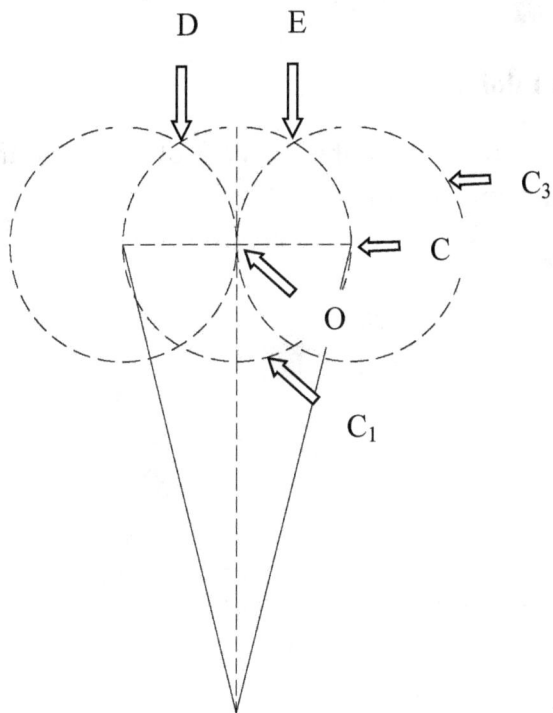

Figure 9 - Mark E, the top intersection of Circle C_1 and C_3.

10 - Mark F the bottom intersection of Circle C_1 and the Bisection Line AO.

What am I doing?

You are marking the center F of the circle that will solve the trisection problem of an angle of 90°.

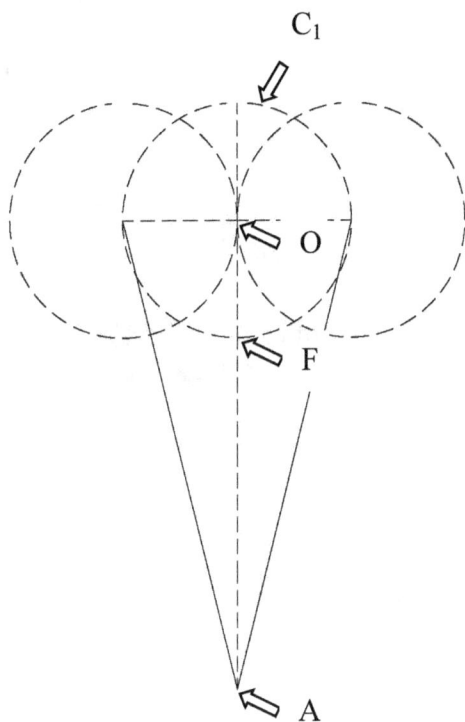

Figure 10 - Mark F the bottom intersection of Circle C_1 and the Bisection Line AO.

11 - Mark G, the intersection of the Bisection Line AO with the bottom part of a Circle with center F and radius FB.

Then draw between the points B and C the top Arc C_4, with center F and radius FB.

What am I doing?

You are drawing the Arc **C_4** that solve the trisection of a 90° angle. You are also marking the center G of the Arc that will solve the trisection problem of an angle of 45°.

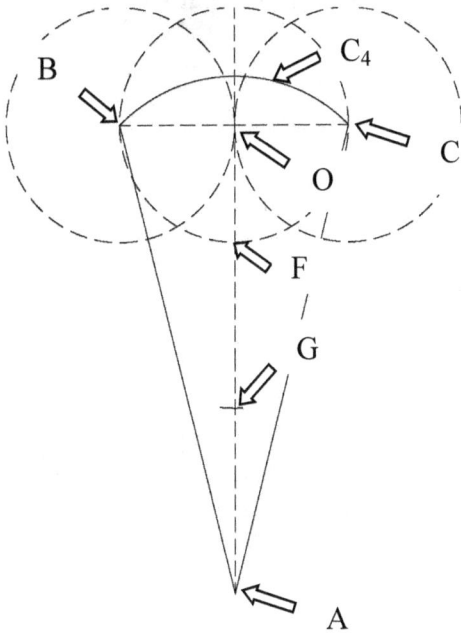

Figure 11 - Mark G, intersection of AO with bottom part of a
Circle C with center F and radius FB

What am I doing?

You are drawing the arc **C₅** that solve the trisection of a 45°
angle.

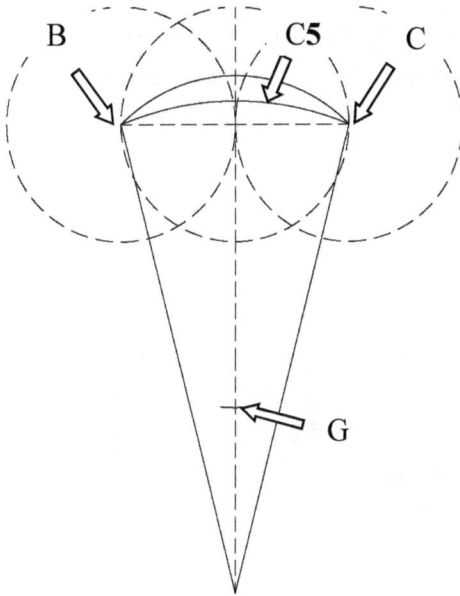

Figure 12 - Draw the top Arc C₅ with center G radius GB

Graphic Trisection of an Arbitrary Angle α
by Harold Florentino LATORTUE, PhD

13 - Mark the point H intersection of a line passing through the points F and D with the Arc C_4.

What am I doing?

You are finding the first point solution for trisecting an angle of 90°. The angle BFH is that solution and its value is 30°.

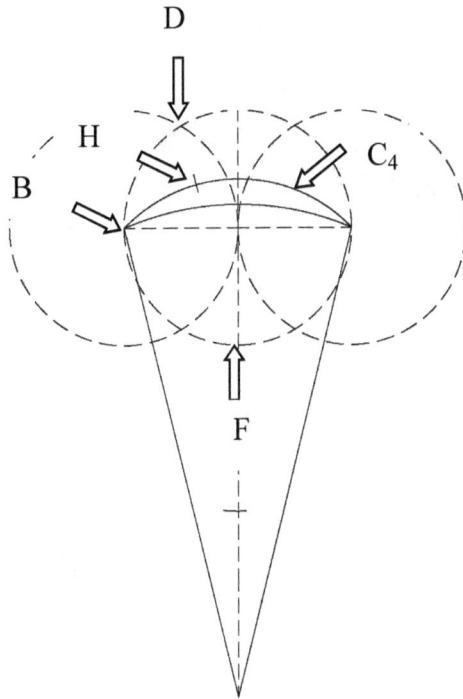

Figure 13 - Mark the point H intersection of line F and D with the Arc C_4.

14 - Mark the point I intersection of a line passing through the points F and E with the Arc C_4.

What am I doing?

You are finding the second point solution for trisecting an angle of 90°. The angles HFI and IFC are with the angle BFH the three angles of the solution of trisecting of an angle α with apex F and equals to 90°. These three angles are all equal to 30°.

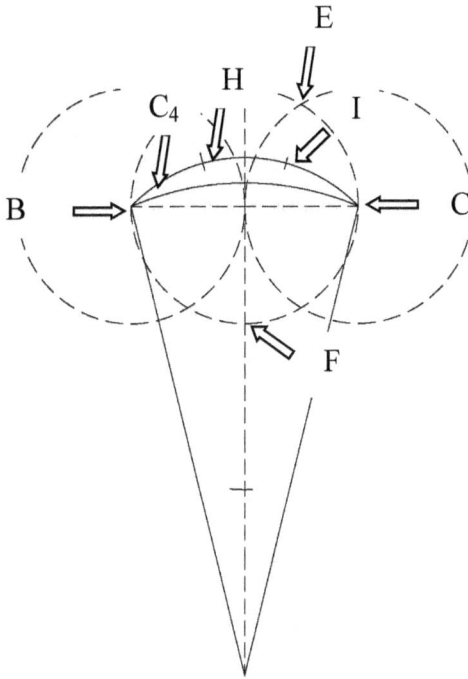

Figure 14 - Mark I intersection of line FE with the Arc C₄.

15 - Mark the point J intersection of the Arc C_5 with a line passing through the points G and H.

What am I doing?

You are finding the first point solution for trisecting an angle of 45°. The angle BGJ is that solution and its value is 15°.

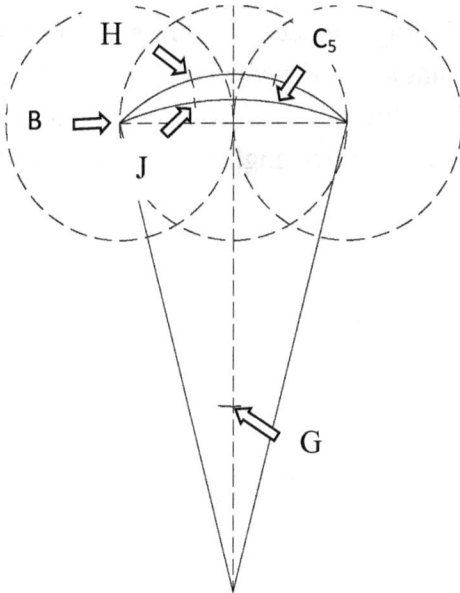

Figure 15 - Mark J intersection of line GH with the Arc C_5.

16 - Mark the point K intersection of the Arc C_5 with a line passing through the points G and I.

What am I doing?

You are finding the second point solution for trisecting an angle of 45°. The angles JGK and KGC are with the angle BGJ the three angles of the solution of trisecting of an angle α with apex G and equals to 45°. These three angles are all equal to 15°.

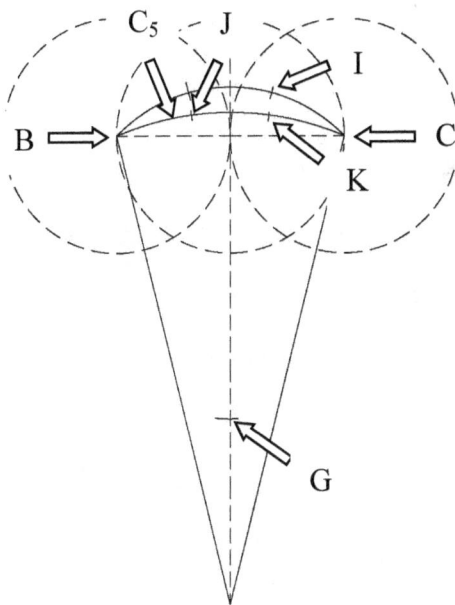

Figure 16 - Mark K intersection of line GI with the Arc C_5.

17 - Mark L the top intersection of Circle C_1 with the bisection Line AO and then mark the point M.

M is the intersection of the bisection Line AO with a circle with center L and radius equals to LB

What am I doing?

You are marking the center M of the Arc that will solve the trisection problem of an angle of 135°.

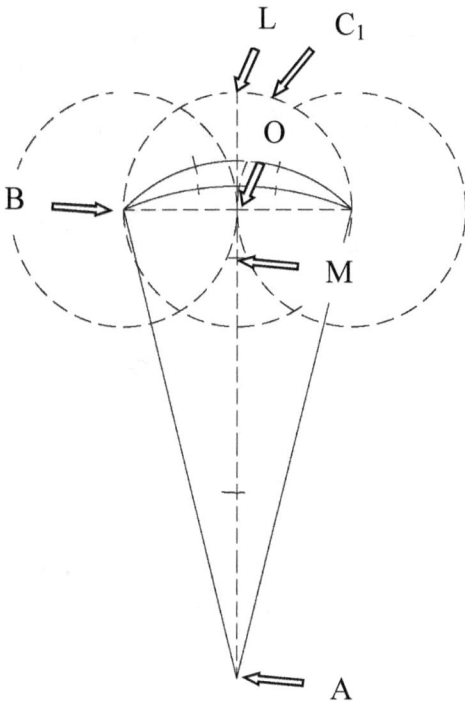

Figure 17 - Mark L top intersection of C_1 with AO and mark M intersection of AO with a circle with center L and radius equals to LB.

Graphic Trisection of an Arbitrary Angle α
by Harold Florentino LATORTUE, PhD

18 - Draw the top Arc C_6 with center M and radius equals to MB.

What am I doing?

You are drawing the arc $\mathbf{C_6}$ that solve the trisection of a 135° angle.

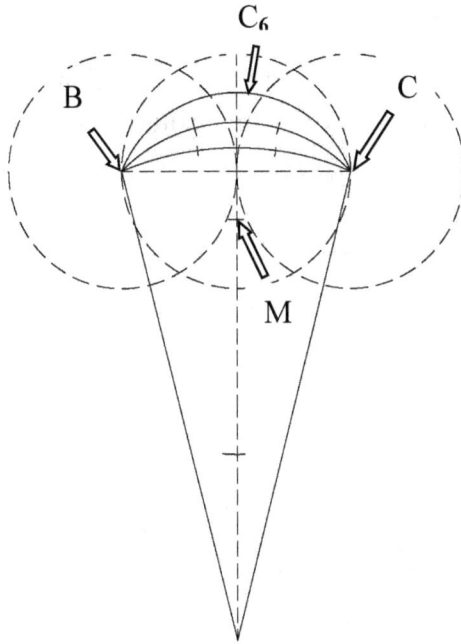

Figure 18 - Draw top Arc C_6 with center M and radius equals to MB

19 - Mark the point N intersection of a line passing through the points L and B with the Arc C_6.

What am I doing?

You are finding the first point solution for trisecting an angle of 135°. The angle BMN is that solution and its value is 45°.

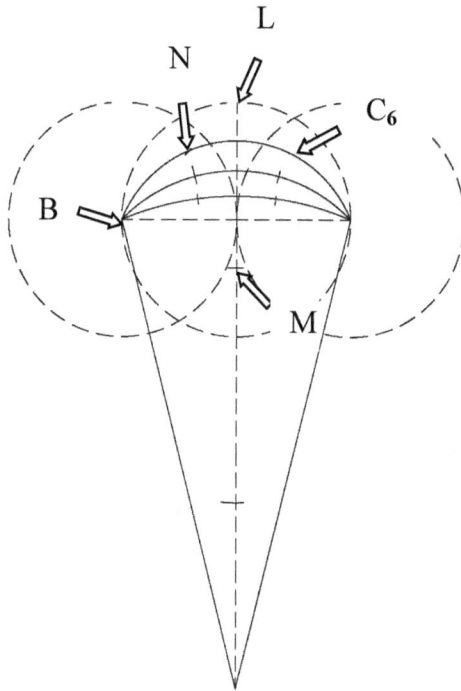

Figure 19 - Mark the point N intersection of a line passing
through the points L and B with the Arc C_6.

20 - Mark the point Q, intersection of the Arc C_6 with a line passing through the points L and C.

What am I doing?

You are finding the second point solution for trisecting an angle of 135°. The angles NMQ and QMC are, with the angle BMN, the three angles of the solution of trisecting of an angle α with apex M and equals to 135°. These three angles are all equal to 45°.

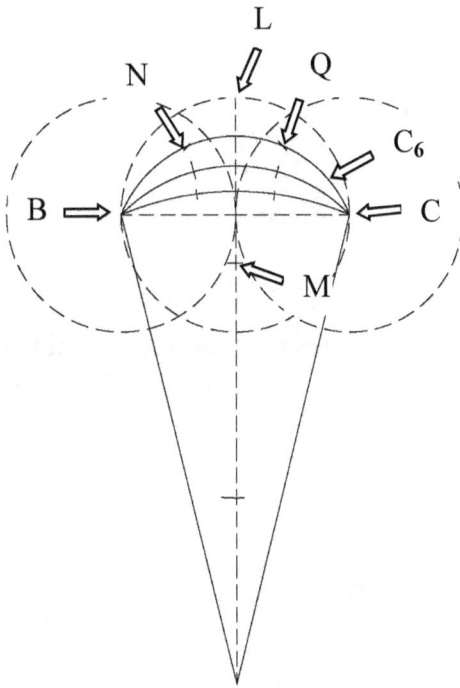

Figure 20 - Mark Q intersection of line LC with the Arc C_6.

21 - Mark the points R and S that trisect the segment BC. (Annex 1)

What am I doing?

You are finding the points solution for trisecting an angle of 0°, with the FLatortue Method. This assertion will look awkward since we all know that the trisection of an angle α of 0° produces three angles of zero-degree value and that both sides of the angles are on just one line. However, when you consider that a zero-degree angle has its apex at an infinity location, the sides then are parallel and cross at that infinity point. Thus, the assertion does make sense.

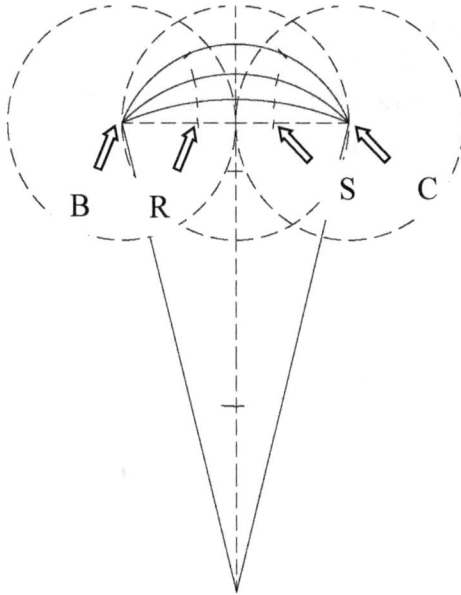

Figure 21 - Mark the points R and S that trisect line BC.

22 - Draw the Arc C_7 passing through the points E, Q and I, then draw the arc C_8 passing through the points I, K and S.

The curve EQIKS forms Locus 1.

What am I doing?

You are drawing the first locus of all point solutions for trisecting an angle α from 0° to 180°.

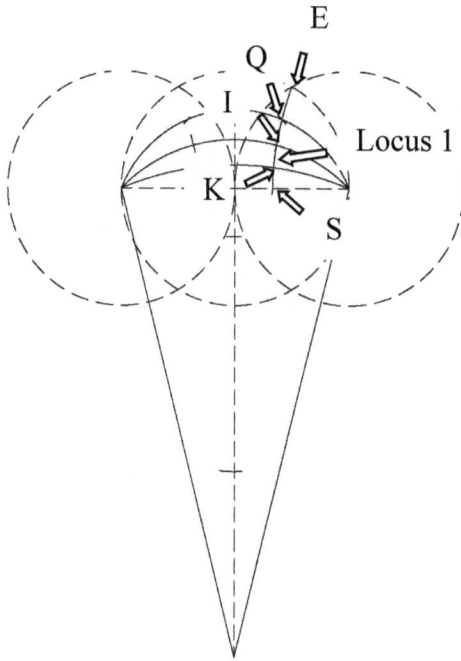

Figure 22 - Draw arc C_7 passing through the points E, Q and I.
Then draw arc C_8 passing through the points I, K and S.

23 - Draw the Arc C_9 passing through the points D, N and H, then draw the arc C_{10} passing through the points H, J and R.

The curve DNHJR forms Locus 2.

What am I doing?

You are drawing the second locus of all point solutions for trisecting an angle α from 0° to 180°.

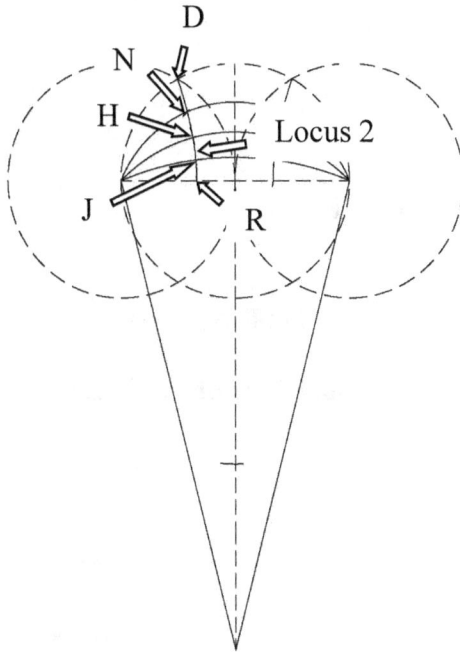

Figure 23 - Draw Arc C_9 through D, N and H, then draw Arc C_{10} through H, J and R.

What must I do?

- Draw the top Arc C_{11} with radius equals AB and centered at the Apex A.

- Mark the points P_1 and P_2, intersections of Arc C_{11} with respectively Locus 1 and Locus 2.

- Draw the lines P_1A and P_2A.

The three equal angles solutions of the Trisection of α are the angles:

BAP_2, P_2AP_1 and P_1AC.

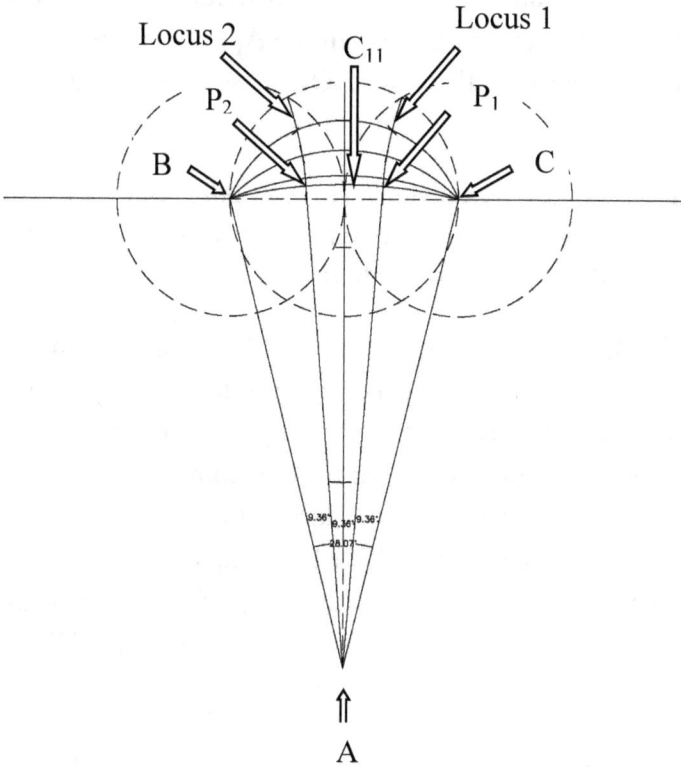

Figure 24 - Solutions of the trisection of an arbitrary angle α.

i - Locus 1 and Locus 2 are the two curves where all points solutions P_1 and P_2 are located when the Apex A of the angle to be trisected is on the bisection line AO. These loci are parts of two Hyperbolas.

ii - The difference between the Algebraic and Graphic solutions was tested for various values of **α**. The results confirm the accuracy of the Flatortue method.

iii - Even though the hyperbola loci will provide all the points solutions from **α** equals to zero to 360°, the Flatortue Method as defined above covers the range of **α** from zero to 180°. However, that is all that is needed for the entire range from zero to 360°. When solving for angle α° larger than 180°, work on the angle $\phi° = 360° - α°$, apply the FLatortue method on $\phi°$. Then subtract the result ($\phi°/3$) from a 120° angle (Annex 2) to get the solutions for the trisection of the angle α° greater than 180°.

iv - For α equals to zero and α equals to 360°, these two angles are defined by only one line and one apex and not by two lines and the apex as are all other angles. However, the solutions for these singular angles are already known since:

a - Dividing a zero angle into three equal parts produces three angles of values all equal zero.

b - Dividing a 360° angle can be easily achieved by constructing three angles of 120° or (180° - 60°).

ALGEBRAIC SOLUTION

Algebraic Approach to Solution

Given an angle BAC with Apex A and whose value is equal to an unknown number α, from the side of the angle A, construct an isosceles triangle ABC with sides AB and AC being equal. Then find the bisector AO of the angle BAC.

At the intersect 'O' of the bisector line and the side BC of the triangle ABC, draw a Cartesian Coordinates System with the X axis going through the line BC and the Y axis passing through the bisector line AO. The origin of the Cartesian coordinates system is the O (0,0).

Draw the top arc BC with center A and radius AB.

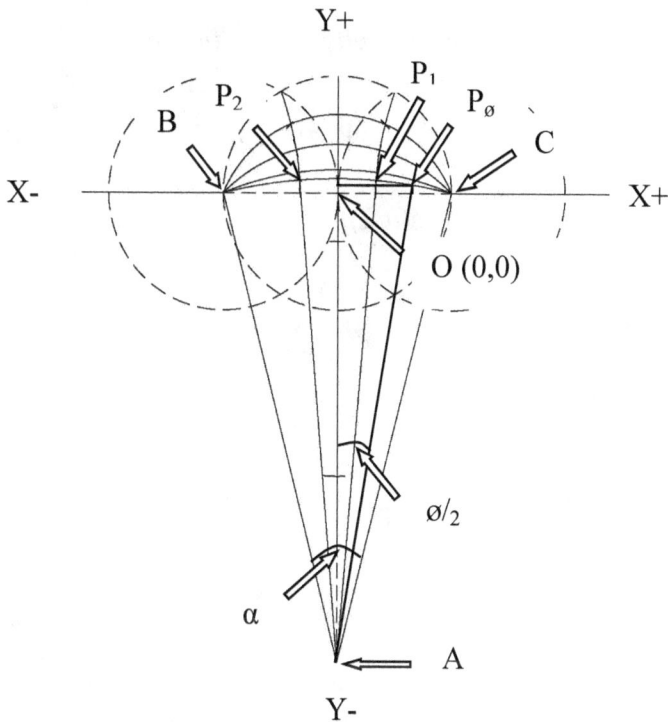

Figure 25 - Cartesian coordinates system

Equations for the algebraic solution:

For the angle α to be divided by three equal parts or trisection, using figure above, the two trisection lines AP_1 and AP_2 will be on the arc BC with center A. The coordinates of any point P_ϕ on the arc including the points P_1 and P_2 can be written as:

P_ϕ [$R\sin(\phi_P)$, $R\cos(\phi_P)$ - $R\cos(\alpha/2)$]

Where:

(ϕ_P) is the value of the angle >PAO

and

(α) is the value of the arbitrary angle to be divided by three or trisected.

and

R is the Radius of the circle centered at the apex A of the angle to be trisected and crossing the X axis at points B and C.

Thus, the coordinates of the points P_1 and P_2, solutions for the trisection of the angle BAC, are:

$Y = R\cos(\alpha/6) - R\cos(\alpha/2)$

Or

$Y = R[\cos(\alpha/6) - \cos(\alpha/2)]$ (1)

and

$X = \pm R\sin(\alpha/6)$ (2)

Form A or Parametric Form of the equation

$X = \pm R\sin(\alpha/6)$

$Y = R(\cos(\alpha/6) - \cos(\alpha/2))$

Form A of the equation

$X = \pm R\sin(\alpha/6)$

$Y = R(\cos(\alpha/6) - \cos(\alpha/2))$

NB: It is important to notice that Form A defines two equations and not one.

a - The first equation defines the point P_1 by $X = +R\sin(\alpha/6)$

and

b - the second equation defines the point P_2 by $X = -R\sin(\alpha/6)$.

In the followings we are going to analyze the first equation whose points are in the first quadrant of the trigonometric circle by fixing the range to the positive value of X.

Form B of the equation

Let's 'K' equals the length of the segment BO, then

$K = R\sin(\alpha/2)$ or $K^2 = R^2\sin^2(\alpha/2)$ (3)

From equations (1), (2) and (3) we have:

$Y^2 = R^2[\cos(\alpha/6) - \cos(\alpha/2)]^2$

$Y^2 = R^2[\cos^2(\alpha/6) + \cos^2(\alpha/2)] - 2 R^2 \cos(\alpha/6) \cos(\alpha/2)$

$Y^2 = R^2[1 - \sin^2(\alpha/6) + 1 - \sin^2(\alpha/2)] - 2 R^2 \cos(\alpha/6) \cos(\alpha/2)$

$Y^2 = R^2 - R^2\sin^2(\alpha/6) + R^2 - R^2\sin^2(\alpha/2) - 2 R^2 \cos(\alpha/6) \cos(\alpha/2)$

$Y^2 = R^2 - X^2 + R^2 - K^2 - 2 R^2\cos(\alpha/6)\cos(\alpha/2)$

$Y^2 + X^2 = 2 R^2 - K^2 - 2 R^2\cos(\alpha/6)\cos(\alpha/2)$

$$Y^2 + X^2 = 2R^2 - K^2 - 2R^2\cos(\alpha/6)\cos(\alpha/2)$$

Since

$Y = R\cos(\alpha/6) - R\cos(\alpha/2)$

$Y/R = \cos(\alpha/6) - \cos(\alpha/2)$

$\cos(\alpha/6) = Y/R + \cos(\alpha/2)$

Thus, we have

$Y^2 + X^2 = 2R^2 - K^2 - 2R^2[(Y/R + \cos(\alpha/2)]\cos(\alpha/2)$

$Y^2 + X^2 = 2R^2 - K^2 - 2RY\cos(\alpha/2) - 2R^2\cos^2(\alpha/2)$

$Y^2 + X^2 = 2R^2(1 - \cos^2(\alpha/2)) - K^2 - 2RY\cos(\alpha/2)$

$Y^2 + X^2 = 2R^2\sin^2(\alpha/2) - K^2 - 2RY\cos(\alpha/2)$

Since $K^2 = R^2\sin^2(\alpha/2)$

then

$Y^2 + X^2 = 2K^2 - K^2 - 2RY\cos(\alpha/2)$

$Y^2 + X^2 = K^2 - 2RY\cos(\alpha/2)$

This equation can be rewritten as:

$Y^2 + 2RY\cos(\alpha/2) + X^2 = K^2$

$[Y + R\cos(\alpha/2)]^2 - R^2\cos^2(\alpha/2)) + X^2 = K^2$

$[Y + R\cos(\alpha/2)]^2 + X^2 = K^2 + R^2\cos^2(\alpha/2))$

$[Y + R\cos(\alpha/2)]^2 + X^2 = K^2 + R^2[1 - \sin^2(\alpha/2)]$

$[Y + R\cos(\alpha/2)]^2 + X^2 = K^2 + R^2 - R^2\sin^2(\alpha/2)$

$[Y + R\cos(\alpha/2)]^2 + X^2 = K^2 + R^2 - K^2$

$[Y + R\cos(\alpha/2)]^2 + X^2 = R^2$

Form B of the equation

$[Y + R\cos(\alpha/2)]^2 + X^2 = R^2$

---

1 - This is the equation of a circle on the form of:

$$(Y - Y_0)^2 + (X - X_0)^2 = R^2$$

With : $Y_0 = -R\cos(\alpha/2)$

$X_0 = 0$

2 - The equation is independent of the segment K.

3 - One can notice that the circle changes both size and position when the angle α varies. The reasons are:

a - $Y_0 = -R\cos(\alpha/2)$ which is not a constant number.

b - by keeping BO fixed, any time you change the value of α, the value of R changes.

4- The center A (Apex) of the circle will slide up or down on the Y axis depending on whether you increase or decrease the value of α.

5- The corresponding arc BC will rise or flatten depending on whether you increase or decrease the value of α.

These characteristic behaviors of the graph are very important to remember. They are the keys that allow someonc to determine the trisection of any arbitrary angle α.

Graphic Trisection of an Arbitrary Angle α
by Harold Florentino LATORTUE, PhD

The algebraic solution:

For any angle BAC of value α, the algebraic procedure for dividing the angle into three equal parts is as follow:

1 - For α find the Radius R

$$R = K/\sin(\alpha/2)$$

2 - Using R, determine X

$$X_1 = R\sin(\alpha/6)$$

$$X_2 = -R\sin(\alpha/6)$$

3 - Then compute Y using

$$Y_1 = Y_2 = R(\cos(\alpha/6) - \cos(\alpha/2))$$

4 - Points P_1 and P_2

$$P_1(X_1, Y_1) \text{ and } P_2(X_2, Y_2)$$

5 - Find the coordinates of the apex A of the angle:

a - $\quad \mathbf{Y_A} = -R\cos(\alpha/2)$

b - $\quad \mathbf{X_A} = 0$

c - $\quad$ A $(0, \mathbf{Y_A})$

The three angles of the trisection are delimited by:

Line AB : A $(0, Y_A)$ and B $(-K, 0)$

Line P_2A : A $(0, Y_A)$ and $P_2(X_2, Y_2)$

Graphic Trisection of an Arbitrary Angle α
by Harold Florentino LATORTUE, PhD

Line P_1A : A $(0, X_A)$ and $P_1(X_1, Y_1)$

Line AC : A $(0, Y_A)$ and C $(+K, 0)$

The three equal angles of the solution are: $>BAP_2$, $> P_2AP_1$ and $>P_1AC$.

Remarks:

1 - To use the algebraic method to trisect an angle α, one must know the arithmetic value of the angle. This value is not necessarily available when the angle is just drawn on a sheet of paper.

2 - Some angles such as 180°, 90° already have known graphic solutions for their trisections.

3 - This proposed algebraic procedure to find the trisection is true for any value of α except for α equals zero and 360°, which produced a value of ∞ for the radius R. However, the solutions for these singular angles are already known since:

a - Dividing a zero angle into three equal parts produces three angles of values equal zero.

b - Dividing a 360° angle can be easily achieved by constructing three angles of 120° (180° - 60°).

Case of α equals 180°

This is one of the angle for which there are easy ways to divide it graphically by three. Let's choose **K equals 1**. (One must remember that K can be of any value).

$$B (-K, 0) \text{ or } B (-1,0)$$

$$C (+K, 0) \text{ or } B (+1,0)$$

1 - For $\alpha = 180°$ find the Radius R

$$R_{180} = 1/\sin(180°/2) = 1/\sin(90°) = 1/1$$

$$\mathbf{R_{180} = 1}$$

2 - Using R_{180}, determine X_{180}

$$X_{P1} = \quad 1*\sin(180°/6) = 1*\sin(30°) = 1*(1/2)$$

$$\mathbf{X_{P1} = \quad +0.5}$$

$$X_{P2} = -1*\sin(180°/6) = -1*\sin(30°) = -1*(1/2)$$

$$\mathbf{X_{P2} = \quad -0.5}$$

3 - Then compute Y_{180} using

$$Y_{P1} = Y_{P2} = R[\cos(180°/6) - \cos(180°/2)]$$

$$Y_{P1} = Y_{P2} = 1[\cos(180°/6) - \cos(180°/2)]$$

$$Y_{P1} = Y_{P2} = 1[\cos(30°) - \cos(90°)]$$

$$Y_{P1} = Y_{P2} = 1 (3^{1/2}/2 - 0)$$

$$\mathbf{Y_{P1} = Y_{P2} = 0.866025}$$

Graphic Trisection of an Arbitrary Angle α
by Harold Florentino LATORTUE, PhD

4 - Points P_1 and P_2

$$P_1(0.5, 0.866025) \text{ and } P_2(-0.5, 0.866025)$$

5 - Find the coordinates of the apex A of the angle:

a - $Y_A = -R\cos(\alpha/2)$

$Y_A = -1*\cos(180º/2) = -1*\cos(90º) = -1*0$

$Y_A = 0$

b - $X_A = 0$

$X_{180} = 0$

c - $A_{180} (0, Y_A)$

$A_{180} (0, 0)$

The three angles of the trisection are delimited by:

Line AB   : $A_{180} (0, 0)$ and $B (-1, 0)$

Line P_1A : $A_{180} (0, 0)$ and $P_1(+0.5, +0.866025)$

Line P_2A : $A_{180} (0, 0)$ and $P_2(-0.5, +0.866025)$

Line AC   : $A_{180} (0, 0)$ and $C (+1, 0)$

The three equal angles of the solution are: BAP_1, P_1AP_2 and P_2AC

Case of α equals 90°

This is one of the angle for which there are easy ways to divide it graphically by three. Let's choose **K equals 1**. (One must remember that K can be of any value). Thus, the coordinates of B and C are:

$$B\ (-K,\ 0)\ \text{or}\ B\ (-1,\ 0)$$

$$C\ (+K,\ 0)\ \text{or}\ B\ (+1,\ 0)$$

1 - For $\alpha = 90°$ find the Radius R

$$R_{90} = 1/\sin(90°/2) = 1/\sin(45°) = 1/(2^{1/2}/2)$$

$$\mathbf{R_{90} = 1.414213}$$

2 - Using R_{90}, determine X_{90}

$$X_{P1} = \ 1.414213\sin(90°/6) = 1.414213\sin(15°)$$

$$X_{P1} = \ 1.414213 \times 0.258819$$

$$\mathbf{X_{P1} = \ +0.366025}$$

$$X_{P2} = -1.414213\sin(90°/6) = 1.414213\sin(15°)$$

$$X_{P2} = \ -1.414213*0.258819$$

$$\mathbf{X_{P2} = \ -0.366025}$$

3 - Then compute Y_{90} using

$$Y_{P1} = Y_{P2} = R[\cos(90°/6) - \cos(90°/2)]$$

$$Y_{P1} = Y_{P2} = 1.414213[\cos(90°/6)-\cos(90°/2)]$$

$$Y_{P1} = Y_{P2} = 1.414213\ [\cos(15°) - \cos(45°)]$$

Graphic Trisection of an Arbitrary Angle α
by Harold Florentino LATORTUE, PhD

$$Y_{P1} = Y_{P2} = 1.414213*(0.258819)$$

$$\mathbf{Y_{P1} = Y_{P2} = +0.366025}$$

4 - Points P_1 and P_2

$$\mathbf{P_1(+0.366025, +0.366025) \text{ and } P_2(-0.366025, + 0.366025)}$$

5 - Find the coordinates of the apex A of the angle:

a - $Y_A = - R\cos(\alpha/2)$

$Y_A = - 1.414213 \cos(90°/2) = - 1.414213 \cos(45°)$

$Y_A = - 1.414213 *0.707106$

$\mathbf{Y_A = - 1}$

b - $X_A = 0$

$\mathbf{X_{90} = 0}$

c - $A_{90} (0, Y_A)$

$\mathbf{A_{90} (0, - 1)}$

The three angles of the trisection are delimited by:

Line AB   : $\mathbf{A_{90} (0, - 1) \text{ and } B(-1, 0)}$

Line P_1A : $\mathbf{A_{90} (0, - 1) \text{ and } P_1(0.366025, 0.366025)}$

Line P_2A : $\mathbf{A_{90} (0, - 1) \text{ and } P_2(-0.366025, 0.366025)}$

Line AC   : $\mathbf{A_{90} (0, - 1) \text{ and } C (+1, 0)}$

The three equal angles of the solution are: BAP_1, P_1AP_2 and P_2AC

Graphic Trisection of an Arbitrary Angle α
by Harold Florentino LATORTUE, PhD

Case of α equals 45°

This is one of the angle for which there are easy ways to divide it graphically by three. Let's choose **K equals 1**. (One must remember that K can be of any value). Thus, the coordinates of B and C are:

$$B(-K, 0) \text{ or } B(-1, 0)$$

$$C(+K, 0) \text{ or } B(+1, 0)$$

1 - For α = 45° find the Radius R

$$R_{45} = 1/\sin(45°/2) = 1/\sin(22.5°) = 1/(0.382683)$$

$R_{45} = 2.613125$

2 - Using R_{45}, determine X_{45}

$$X_{P1} = 2.613125\sin(45°/6) = 2.613125\sin(7.5°)$$

$$X_{P1} = 2.613125*0.130526$$

$X_{P1} = +0.341081$

$$X_{P2} = -2.613125\sin(45°/6) = -2.613125\sin(7.5°)$$

$$X_{P2} = -2.613125*0.130526$$

$X_{P2} = -0.341081$

3 - Then compute Y_{45} using

$$Y_{P1} = Y_{P2} = R[\cos(45°/6) - \cos(45°/2)]$$

$$Y_{P1} = Y_{P2} = 2.613125*[\cos(45°/6) - \cos(45°/2)]$$

Graphic Trisection of an Arbitrary Angle α
by Harold Florentino LATORTUE, PhD

$$Y_{P1} = Y_{P2} = 2.613125 \, [\cos(7.5°) - \cos(22.5°)]$$

$$Y_{P1} = Y_{P2} = 2.613125*(0.991444 - 0.923879)$$

$$Y_{P1} = Y_{P2} = 2.613125 *0.067565$$

$$\mathbf{Y_{P1} = Y_{P2} = 0.176556}$$

4 - Points P_1 and P_2

P_1 (0.341081, 0.176556) and P_2 (-0.341081, 0.176556)

5 - Find the coordinates of the apex A of the angle:

a - $\quad Y_A = - R\cos(\alpha/2)$

$\quad\quad Y_A = -2.613125\cos(45°/2) = -2.613125\cos(22.5°)$

$\quad\quad Y_A = - 2.613125*0.923879$

$\quad\quad \mathbf{Y_A = - 2.414213}$

b - $\quad \mathbf{X_A = 0}$

$\quad\quad \mathbf{X_{45} = 0}$

c - $\quad A_{45}$ (0, Y_A)

$\quad\quad \mathbf{A_{45}\ (0, - 2.414213)}$

The three angles of the trisection are delimited by:

Line AB : $\mathbf{A_{45}}$ **(0, - 2.414213) and B (-1, 0)**

Line P_1A : $\mathbf{A_{45}}$ **(0, - 2.414213) and P₁(0.341081, 0.176556)**

Line P_2A : $\mathbf{A_{45}}$ **(0, - 2.414213) andP₂(-0.341081, 0.176556)**

Graphic Trisection of an Arbitrary Angle α
by Harold Florentino LATORTUE, PhD

Line AC: A_{45} (0, - 2.414213) and C (+1, 0)

The three equal angles of the solution are: BAP_1, P_1AP_2 and P_2AC

Case of α equals 135°

This is one of the angle for which there are easy ways to divide it graphically by three. Let's choose **K equals 1**. (One must remember that K can be of any value). Thus, the coordinates of B and C are:

$$B (-K, 0) \text{ or } B (-1, 0)$$

$$C (+K, 0) \text{ or } B (+1, 0)$$

1 - For α = 135° find the Radius R

$$R_{135} = 1/\sin(135°/2) = 1/\sin(67.5°) = 1/(0.923879)$$

$$\mathbf{R_{135} = 1.082392}$$

2 - Using R_{135}, determine X_{135}

$$X_{P1} = 1.082392\sin(135°/6) = 1.082392\sin(22.5°)$$

$$X_{P1} = 1.082392*0.382683$$

$$\mathbf{X_{P1} = + 0.414213}$$

$$X_{P2} = - 1.082392\sin(135°/6) = - 1.082392\sin(22.5°)$$

$$X_{P2} = - 1.082392*0.382683$$

$$\mathbf{X_{P2} = - 0.414213}$$

3 - Then compute Y_{45} using

$$Y_{P1} = Y_{P2} = R[\cos(135°/6) - \cos(135°/2)]$$

$$Y_{P1} = Y_{P2} = 1.082392 [\cos(135°/6) - \cos(135°/2)]$$

$$Y_{P1} = Y_{P2} = 1.082392 \, (\cos(22.5^\circ) - \cos(67.5^\circ))$$

$$Y_{P1} = Y_{P2} = 1.082392 \, (0.923880 - 0.382683)$$

$$Y_{P1} = Y_{P2} = 1.082392 * 0.541197$$

$$\mathbf{Y_{P1} = Y_{P2} = 0.585786}$$

4 - Points P_1 and P_2

$P_1(0.414213, 0.585786)$ and $P_2(- 0.414213, 0.585786)$

5 - Find the coordinates of the apex A of the angle:

a - $Y_A = - R\cos(\alpha/2)$

$$Y_A = -1.082392\cos(135^\circ/2) = -1.082392\cos(67.5^\circ)$$

$$Y_A = - 1.082392 * 0.382683$$

$$\mathbf{Y_A = - 0.414214}$$

b - $\mathbf{X_A = 0}$

$$\mathbf{X_{135} = 0}$$

c - $A_{135} \, (0, \, Y_A)$

$$\mathbf{A_{135} \, (0, \, - 0.414214)}$$

The three angles of the trisection are delimited by:

Line AB : $\mathbf{A_{135}(0, - 0.414214)}$ and **B $(-1, 0)$**

Line P_1A : $\mathbf{A_{135}(0, -0.414214)}$ and $\mathbf{P_1(0.414213, 0.585786)}$

Line P_2A : $\mathbf{A_{135}(0,-0.414214)}$ and $\mathbf{P_2(-0.414213, 0.585786)}$

Line AC : $\mathbf{A_{135}}$ **(0, - 0.414214)** and **C (+1, 0)**

The three equal angles of the solution are: BAP_1, P_1AP_2 and P_2AC

Table of algebraic solution for various values of α°

α°	K	R	X_{P1}	X_{P2}	$Y_{P1} = Y_{P2}$	X_A	Y_A
180	1	1.0000	0.5000	-0.5000	0.8660	0.0000	0.0000
150	1	1.0353	0.4375	-0.4375	0.6703	0.0000	-0.2679
135	1	1.0824	0.4142	-0.4142	0.5858	0.0000	-0.4142
120	1	1.1547	0.3949	-0.3949	0.5077	0.0000	-0.5774
90	1	1.4142	0.3660	-0.3660	0.3660	0.0000	-1.0000
60	1	2.0000	0.3473	-0.3473	0.2376	0.0000	-1.7321
50	1	2.3662	0.3429	-0.3429	0.1967	0.0000	-2.1445
45	1	2.6131	0.3411	-0.3411	0.1766	0.0000	-2.4142
30	1	3.8637	0.3367	-0.3367	0.1169	0.0000	-3.7321
20	1	5.7588	0.3348	-0.3348	0.0777	0.0000	-5.6713
15	1	7.6613	0.3342	-0.3342	0.0583	0.0000	-7.5958
10	1	11.4737	0.3337	-0.3337	0.0388	0.0000	-11.4301
5	1	22.9256	0.3334	-0.3334	0.0194	0.0000	-22.9038
1	1	114.5930	0.3333	-0.3333	0.0039	0.0000	-114.5887
1.E-06	1	1.E+08	0.3333	-0.3333	0.0000	0.0000	-1.E+08

Table 1 - Table of algebraic solution for various values of α°

α°	Algebraic		Graphic		Difference		Graphic	Diff. $^\circ$
	X_{P1}	Y_{P1}	X_{P1}	Y_{P1}	X_{P1}	Y_{P1}	α°_g	$\alpha^\circ/3 - \alpha^\circ_g$
180	0.50	0.87	0.50	0.87	0.00	0.00	60.0	0.0
150	0.44	0.67	0.44	0.67	0.00	0.00	50.0	0.0
135	0.41	0.59	0.41	0.59	0.00	0.00	45.0	0.0
120	0.39	0.51	0.40	0.51	0.00	0.00	40.0	0.0
90	0.37	0.37	0.37	0.37	0.00	0.00	30.0	0.0
60	0.35	0.24	0.35	0.24	0.00	0.00	20.0	0.0
50	0.34	0.21	0.34	0.20	0.00	0.01	16.7	0.0
45	0.34	0.18	0.34	0.18	0.00	0.00	15.0	0.0
30	0.34	0.12	0.34	0.12	0.00	0.00	10.0	0.0
20	0.33	0.08	0.34	0.08	0.00	0.00	6.7	0.0
15	0.33	0.05	0.33	0.06	0.00	0.00	5.0	0.0
10	0.33	0.04	0.33	0.04	0.00	0.00	3.3	0.0
5	0.33	0.02	0.33	0.02	0.00	0.00	1.7	0.0
1	0.33	0.00	0.33	0.00	0.00	0.00	0.3	0.0
1.00E-06	0.33	0.00	0.33	0.00	0.00	0.00	0.0	0.0

Table 2 - Table of comparison algebraic vs. Graphic solutions

Finding the equation of the first Locus of solution points located in the first quadrant.

Form A of the equation

$X = +R\sin(\alpha/6)$ (1)

$Y = R(\cos(\alpha/6) - \cos(\alpha/2))$ (2)

$K = R\sin(\alpha/2)$ (3)

$Y^2 + X^2 = 2R^2 - K^2 - 2R^2\cos(\alpha/6)\cos(\alpha/2)$ (4)

Rewriting (4)

$Y^2 + K^2 = 2R^2 - 2R^2\cos(\alpha/6)\cos(\alpha/2) - X^2$

Subtracting: $3X^2 + 2KX$ from both side of the equation

$Y^2 + K^2 - (3X^2 + 2KX) = 2R^2 - 2R^2\cos(\alpha/6)\cos(\alpha/2) - X^2 - (3X^2 + 2KX)$

$Y^2 + K^2 - 3X^2 - 2KX = 2R^2 - 2R^2\cos(\alpha/6)\cos(\alpha/2) - X^2 - 3X^2 - 2KX$

$Y^2 + K^2 - 3X^2 - 2KX = 2R^2 - 2R^2\cos(\alpha/6)\cos(\alpha/2) - 4X^2 - 2KX$

Using (1)

$Y^2 + K^2 - 3X^2 - 2KX = 2R^2 - 2R^2\cos(\alpha/6)\cos(\alpha/2) - 4R^2\sin^2(\alpha/6)$

$\qquad\qquad -KR\sin(\alpha/6)$

Using (3)

$Y^2 + K^2 - 3X^2 - 2KX = 2R^2 - 2R^2\cos(\alpha/6)\cos(\alpha/2) - 4R^2\sin^2(\alpha/6)$

$\qquad\qquad - 2R^2\sin(\alpha/2)\sin(\alpha/6)$

$$Y^2 + K^2 - 3X^2 - 2KX = 2R^2[1 - \cos(\alpha/6)\cos(\alpha/2) - 2\sin^2(\alpha/6)$$

$$- \sin(\alpha/2)\sin(\alpha/6)]$$

Since: $1 = \sin^2(\alpha/6) + \cos^2(\alpha/6)$

$$Y^2 + K^2 - 3X^2 - 2KX = 2R^2[\sin^2(\alpha/6) + \cos^2(\alpha/6) - \cos(\alpha/6)\cos(\alpha/2)$$

$$-2\sin^2(\alpha/6) - \sin(\alpha/2)\sin(\alpha/6)]$$

Since:

$$\cos(\alpha/6)\cos(\alpha/2) + \sin(\alpha/2)\sin(\alpha/6) = \cos(\alpha/2 - \alpha/6)$$

$$\cos(\alpha/6)\cos(\alpha/2) + \sin(\alpha/2)\sin(\alpha/6) = \cos(3\alpha/6 - \alpha/6) = \cos(\alpha/3)$$

$$Y^2 + K^2 - 3X^2 - 2KX = 2R^2[\sin^2(\alpha/6) + \cos^2(\alpha/6) - 2\sin^2(\alpha/6) - \cos(\alpha/3)]$$

$$Y^2 + K^2 - 3X^2 - 2KX = 2R^2[\cos^2(\alpha/6) - \sin^2(\alpha/6) - \cos(\alpha/3)]$$

Since:

$$\cos(\alpha/3) = \cos^2(\alpha/6) - \sin^2(\alpha/6)$$

$$Y^2 + K^2 - 3X^2 - 2KX = 2R^2[\cos(\alpha/3) - \cos(\alpha/3)]$$

$$Y^2 + K^2 - 3X^2 - 2KX = 2R^2[0]$$

$$Y^2 + K^2 - 3X^2 - 2KX = 0$$

$$3X^2 + 2KX - Y2 - K^2 = 0 \qquad\qquad (5)$$

Multiplying by 3

$$9X^2 + 6KX - 3Y2 - 3K^2 = 0$$

$$9X^2 + 6KX + K^2 - K^2 - 3Y^2 - 3K^2 = 0$$

Graphic Trisection of an Arbitrary Angle α
by Harold Florentino LATORTUE, PhD Page 86

$(3X + K)^2 - K^2 - 3Y^2 - 3K^2 = 0$

$(3X + K)^2 - K^2 - 3Y^2 - 3K^2 = 0$

$(3X + K)^2 - 3Y^2 = 4K^2$

$9(X + K/3)^2 - 3Y^2 = 4K^2$

$(X + K/3)^2 - Y^2/3 = 4K^2/9$

Dividing by $4K^2/9$

$[(X + K/3)^2 / 4K^2/9] - [(Y^2/3) / 4K^2/9] = 1$

$[(X + K/3)^2 / (2K/3)^2] - [Y^2 / (2(3)^{1/2}K/3)^2] = 1$

This is the equation of a hyperbola such as:

$(X-X_0)^2/a^2 - (Y-Y_0)^2/b^2 = 1$

$X_0 = - K/3$

$Y_0 = 0$

$a^2 = (2K/3)^2$

$b^2 = (2(3)^{1/2}K/3)^2$

Centered at:

$X_c = -K/3$

$Y_c = 0$

Graphic Trisection of an Arbitrary Angle α
by Harold Florentino LATORTUE, PhD

Hyperbola 1

$$(X-X_0)^2/a^2 - (Y-Y_0)^2/b^2 = 1$$

$$X_0 = -K/3$$

$$Y_0 = 0$$

$$a^2 = (2K/3)^2$$

$$b^2 = (2(3)^{1/2}K/3)^2$$

$$X_c = -K/3$$

$$Y_c = 0$$

Finding the equation of second Locus of solution points located in the second quadrant.

Form A of the equation

$X = -R\sin(\alpha/6)$ (1)

$Y = R(\cos(\alpha/6) - \cos(\alpha/2))$ (2)

$K = R\sin(\alpha/2)$ (3)

$Y^2 + X^2 = 2R^2 - K^2 - 2R^2\cos(\alpha/6)\cos(\alpha/2)$ (4)

Rewriting (4)

$Y^2 + K^2 = 2R^2 - 2R^2\cos(\alpha/6)\cos(\alpha/2) - X^2$

Subtracting: $3X^2 - 2KX$ from both side of the equation

$Y^2 + K^2 - (3X^2 - 2KX) = 2R^2 - 2R^2\cos(\alpha/6)\cos(\alpha/2) - X^2 - (3X^2 - 2KX)$

$Y^2 + K^2 - 3X^2 + 2KX = 2R^2 - 2R^2\cos(\alpha/6)\cos(\alpha/2) - X^2 - 3X^2 + 2KX$

$Y^2 + K^2 - 3X^2 + 2KX = 2R^2 - 2R^2\cos(\alpha/6)\cos(\alpha/2) - 4X^2 + 2KX$

Using (1)

$Y^2 + K^2 - 3X^2 + 2KX = 2R^2 - 2R^2\cos(\alpha/6)\cos(\alpha/2)$

$\qquad\qquad -4R^2\sin^2(\alpha/6) - 2KR\sin(\alpha/6)$

Using (3)

$Y^2 + K^2 - 3X^2 + 2KX = 2R^2 - 2R^2\cos(\alpha/6)\cos(\alpha/2)$

$\qquad\qquad - 4R^2\sin^2(\alpha/6) - 2R^2\sin(\alpha/2)\sin(\alpha/6)$

$$Y^2 + K^2 - 3X^2 + 2KX = 2R^2[1 - \cos(\alpha/6)\cos(\alpha/2)$$
$$- 2\sin^2(\alpha/6) - \sin(\alpha/2)\sin(\alpha/6)]$$

Since: $1 = \sin^2(\alpha/6) + \cos^2(\alpha/6)$

$$Y^2+K^2-3X^2+2KX = 2R^2[\sin^2(\alpha/6)+\cos^2(\alpha/6)-\cos(\alpha/6)\cos(\alpha/2)$$
$$-2\sin^2(\alpha/6)- \sin(\alpha/2)\sin(\alpha/6)]$$

Since:

$$\cos(\alpha/6)\cos(\alpha/2) + \sin(\alpha/2)\sin(\alpha/6) = \cos(\alpha/2-\alpha/6)$$

$$\cos(\alpha/6)\cos(\alpha/2) + \sin(\alpha/2)\sin(\alpha/6) = \cos(3\alpha/6-\alpha/6) = \cos(\alpha/3)$$

$$Y^2+K^2-3X^2+2KX=2R^2[\sin^2(\alpha/6) + \cos^2(\alpha/6) - 2\sin^2(\alpha/6) - \cos(\alpha/3)]$$

$$Y^2 + K^2 - 3X^2 + 2KX = 2R^2[\cos^2(\alpha/6) - \sin^2(\alpha/6) - \cos(\alpha/3)]$$

Since:

$$\cos(\alpha/3) = \cos^2(\alpha/6) - \sin^2(\alpha/6)$$

$$Y^2+K^2-3X^2+2KX = 2R^2[\cos^2(\alpha/6) - \sin^2(\alpha/6) - \cos^2(\alpha/6)+ \sin^2(\alpha/6)]$$

$$Y^2 + K^2 - 3X^2 + 2KX = 2R^2[0]$$

$$Y^2 + K^2 - 3X^2 + 2KX = 0$$

$3X^2 - 2KX - Y2 - K^2 = 0$ **(5)**

Multiplying by 3

$$9X^2 - 6KX - 3Y2 - 3K^2 = 0$$

$$9X^2 - 6KX + K^2 - K^2 - 3Y^2 - 3K^2 = 0$$

$(3X - K)^2 - K^2 - 3Y^2 - 3K^2 = 0$

$(3X - K)^2 - 3Y^2 = 4K^2$

$9(X - K/3)^2 - 3Y^2 = 4K^2$

$(X - K/3)^2 - Y^2/3 = 4K^2/9$

Dividing by $4K^2/9$

$[(X - K/3)^2 / 4K^2/9] - [(Y^2/3) / 4K^2/9] = 1$

$[(X - K/3)^2 / (2K/3)^2] - [Y^2 / (2(3)^{1/2}K/3)^2] = 1$

This is the equation of a hyperbola such as:

$(X-X_0)^2/a^2 - (Y-Y_0)^2/b^2 = 1$

$X_0 = + K/3$

$Y_0 = 0$

$a^2 = (2K/3)^2$

$b^2 = (2(3)^{1/2}K/3)^2$

Centered at:

$X_c = +K/3$

$Y_c = 0$

Graphic Trisection of an Arbitrary Angle α
by Harold Florentino LATORTUE, PhD

Hyperbola 2

$(X-X_0)^2/a^2 - (Y-Y_0)^2/b^2 = 1$

$X_0 = +K/3$

$Y_0 = 0$

$a^2 = (2K/3)^2$

$b^2 = (2(3)^{1/2}K/3)^2$

$X_c = +K/3$

$Y_c = 0$

Sketch of algebraic solution for points in the first quadrant for various values of α^o

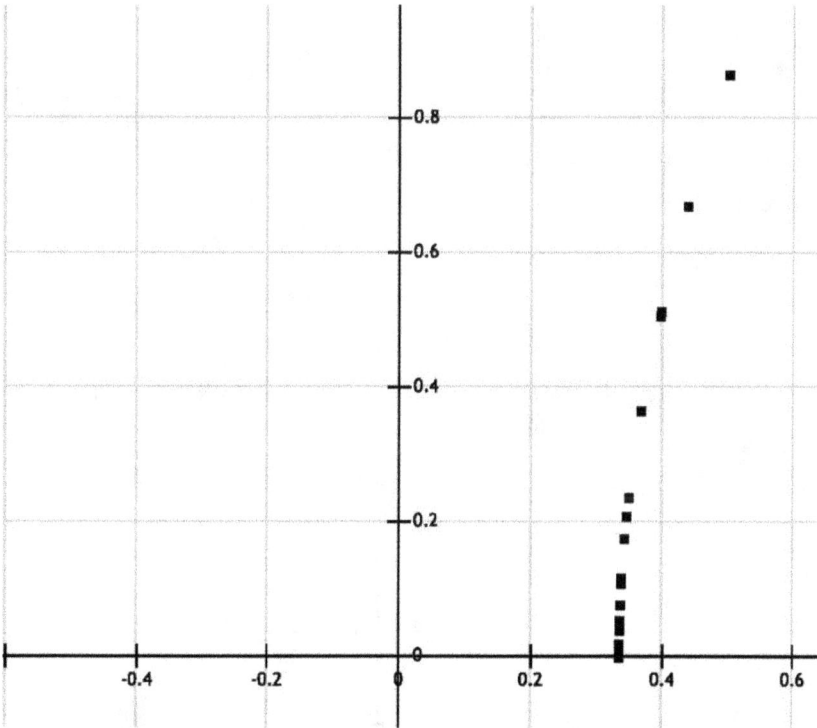

Figure 26 - Sketch of algebraic solution for points in the first quadrant for various values of α^o

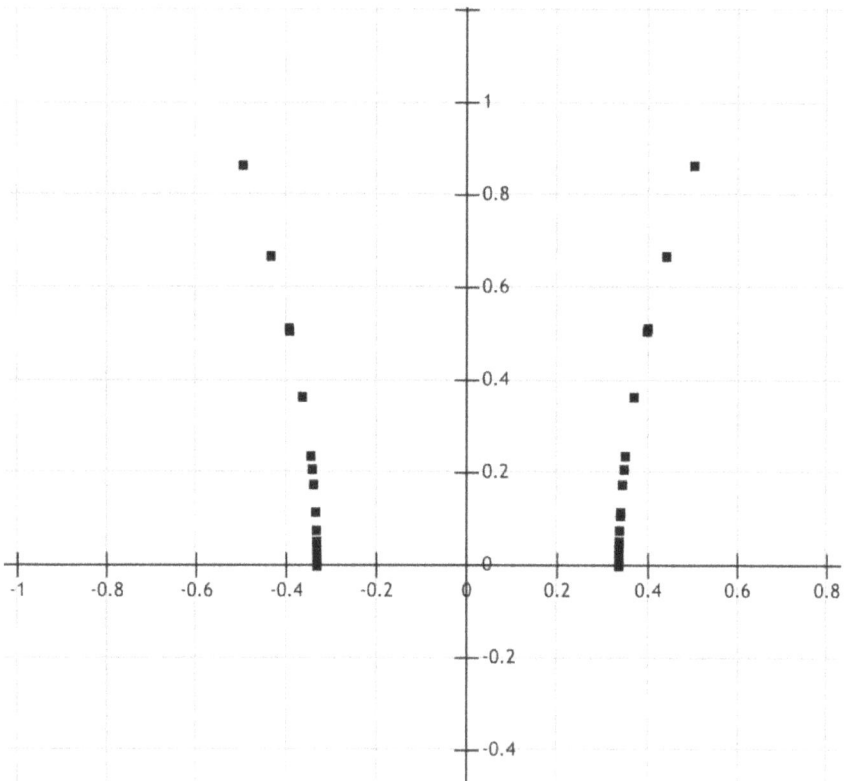

Figure 27 - Sketch of algebraic solution for points in the first and second quadrant for various values of α^o

Sketch of Hyperbola 1 (Details view)

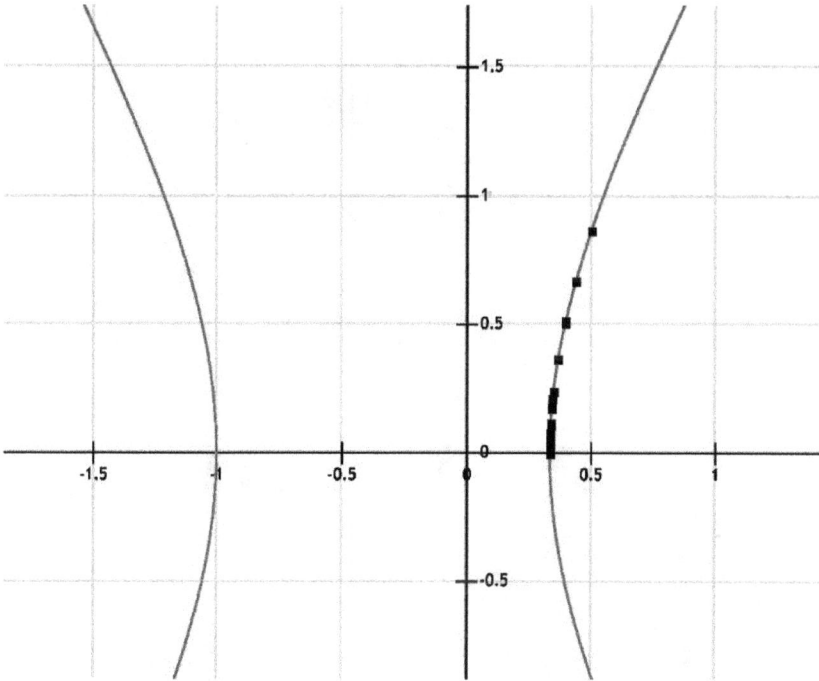

Figure 28 - Sketch of Sketch of Hyperbola 1 (Details view)

Sketch of Hyperbola 1 (whole view)

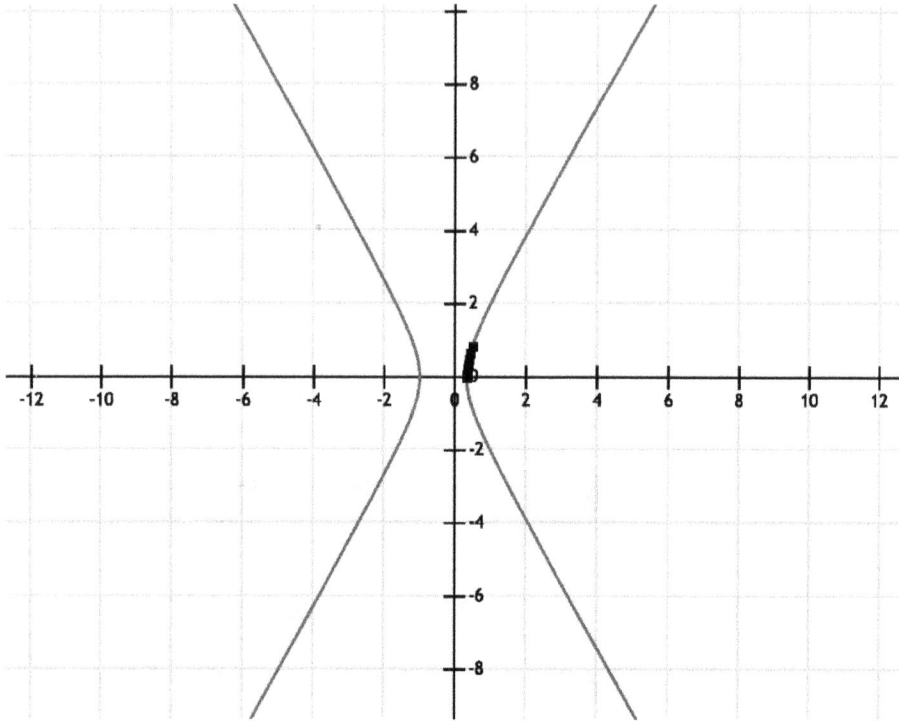

Figure 29 - Sketch of Sketch of Hyperbola 1 (whole view)

Sketch of hyperbolas 1 and 2 with asymptotes

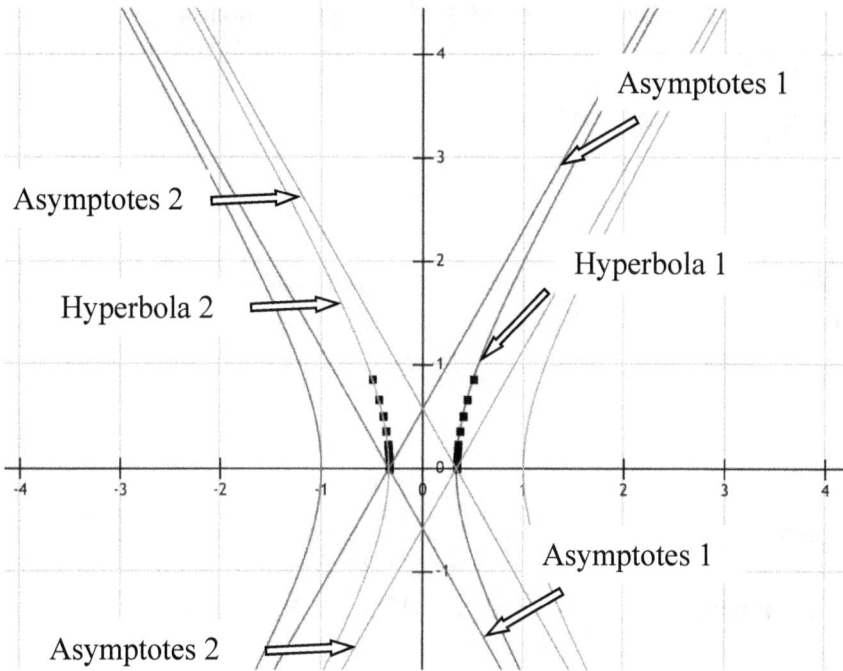

Figure 30 - Sketch of hyperbolas 1 and 2 with asymptotes

Important:

One must notice that the point solutions on the second quadrant are on a different hyperbola than the point solutions on the first quadrant. The analysis of this phenomenon revealed that the points on the first quadrant is on the hyperbola 1 defined by:

$$(X-X_0)^2/a^2 - (Y-Y_0)^2/b^2 = 1$$

with

$$X_0 = -K/3$$

$$Y_0 = 0$$

$$a = 2K/3$$

and

$$b = 2K\,(3)^{1/2}/3$$

This hyperbola is centered at $X = -K/3$ and $Y = 0$

While, the point solutions on the second quadrant are on a different hyperbola defined by:

$$(X-X_0)^2/a^2 - (Y-Y_0)^2/b^2 = 1$$

with

$$X_0 = K/3$$

$$Y_0 = 0$$

$$a = 2K/3$$

and

$b = 2K (3)^{1/2}/3$

This hyperbola 2 is centered at $X = K/3$ and $Y = 0$

Another important remark is the fact that one of the apexes of the hyperbola 1 and one of the apexes of the hyperbola 2 are located on the X axis at + K/3 and -K/3. These points are the solutions for α equals to zero. This confirms that the solution for α equals to zero, in the graphic solution, is justified when set at the trisection point of the segment BC. Certainly, this method will reveal many more characteristics that can be exploited in the future, but this search is beyond the objective of this study.

Conclusions on trisecting an angle α

As one can see, the algebraic method to trisect an angle α is very time consuming, involves many computations and transfer of data into the graph even for very well-known angles α such the 180° and the 90°. Opportunities for errors are important. That is the reason it is necessary to define a graphic way to trisect the angle α.

The objective of this study was to demonstrate that a graphical solution is possible using a compass and a straight edge as it is specified in the problem statement. This objective is achieved successfully. Anyone with basic geometry knowledge can and will be able to trisect any angle α using this rather simple FLatortue Method.

Why this problem was classified as 'impossible' for centuries is beyond understanding? However, by providing the FLatortue method, it is expected that it will be taught in all geometry class levels, while insisting that for centuries it was classified as 'impossible to solve'. The hope is that somewhere on the planet, some smart kid will be inspired and provides solutions to the other problems in the category of 'impossible to solve problems'.

ANNEXES

Annex 1 - How to construct the trisection of the segment BC.

1 - From L draw a first circle with radius equals to LF and mark T, the top intersection of this circle with the line FL.

2 - From T draw a second circle with radius equals to TL.

3 - Draw the line U and V linking the intersections of the two circles.

4 - Mark W the intersection of UV with TB.

5 - Mark R and S the intersections of the line BC with a circle centered at L and radius equals to LW.

6 - The segments BR, RS and SC are the trisections of the segment BC.

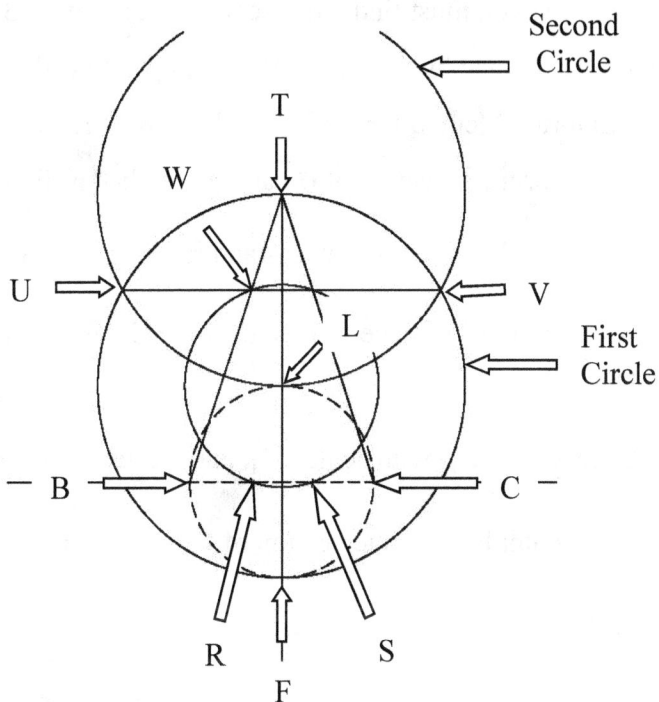

Figure 31 - Construct the trisection of a segment BC.

Annex 2 – Solution for α larger than 180°.

First you must find the trisection angles of $\phi = 360 - \alpha$, which are defined by the angles $BAP_{2\phi}$, $P_{2\phi}AP_{1\phi}$ and $P_{1\phi}AC$ using the FLatortue Method for angle less than 180°. The angles $\alpha/3$ solutions for the trisection of α can be sketched as follow:

1 - Draw Circle C_{12} with center A and radius equals to AB.

2 – Mark point $P'_{1\phi}$ image of $P_{1\phi}$ on Circle C_{12} with respect to the center A.

3 - Draw Circle C_{13} with center $P'_{1\phi}$ and radius equals to $P'_{1\phi}A$.

4 - Mark point $P_{1\alpha}$ the intersection of Circle C_{12} and Circle C_{13}.

5 – Draw line $AP_{1\alpha}$.

6 – Repeat 1 to 5 for point $P_{2\phi}$ to draw line $AP_{1\alpha}$. (The steps for $P_{2\phi}$ are not sketched in the following figure).

The angle $BAP_{2\alpha}$, $P_{2\alpha}AP_{1\alpha}$ and $P_{1\alpha}AC$ are the solution angles for the trisection of α.

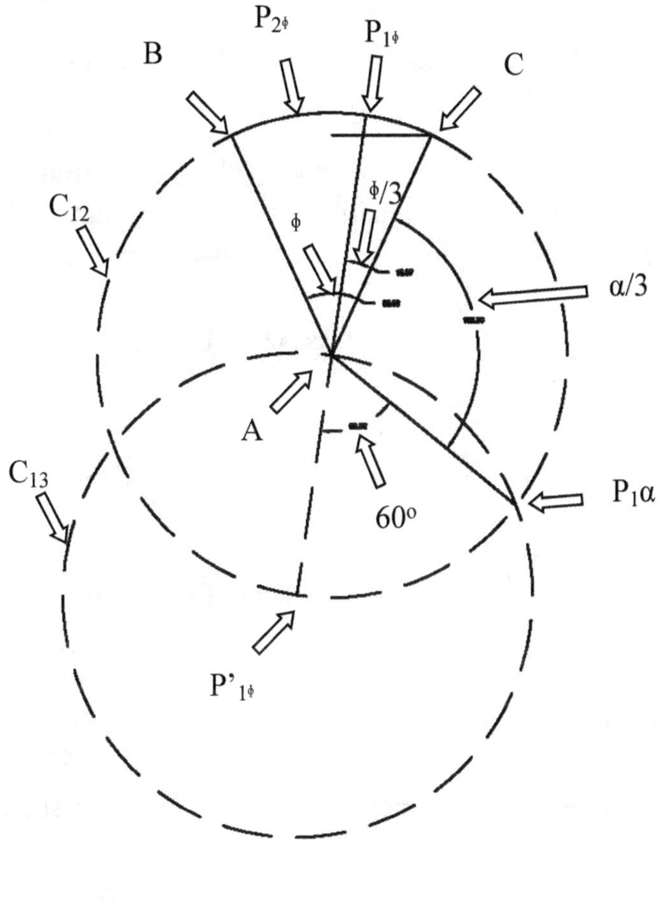

Figure 32 - Solution for α larger than 180°.

Biography

Born in 1953 in Port-au-Prince, Haiti, Dr Harold Florentino Latortue holds several diplomas including:

-   Bachelor Degree in Engineering (July 1977) from the 'Universite d'Etat d'Haiti', Port-au-Prince Haiti
-   Master of Science Degree (Dec 1984) from Texas A&M University, College Station Texas
-   PhD (Dec 1986) from Texas A&M University, College Station Texas.

Dr Latortue has a long experience in both the Private Sector and Public Administration. He served at several high level positions such as:

-   o Advisor to the President of Haiti (2012)
-   o Advisor to the Prime Minister of Haiti (2011)
-   o Secretary of States for Tourism, (2005)
-   o Director General of Ministry of Tourism (2004)
-   o Cabinet Member at Ministry of Trade (2004)
-   o Cabinet Member at Ministry of Tourism (2004)
-   o Director of Cabinet at Ministry of Transportation and Public Works (1993 and 1998)
-   o Director of Natural Resources at Ministry of Agriculture (1987)
-   o Advisor to the Board of Trustees at SocaBank
-   o Member of the Board of Trustees at Union School, Haiti
-   o Advisor to the Director General of Electricity of Haiti (1991)

Dr Latortue speaks English, French, Spanish and Haitian Creole.

Contents

Graphic Trisection of an Arbitrary Angle α

Graphic Trisection of an Arbitrary Angle α

by Harold Florentino LATORTUE, PhD   Page 109

Table of Figures

www.ingramcontent.com/pod-product-compliance
Lightning Source LLC
Chambersburg PA
CBHW052109230326
41599CB00054B/5267